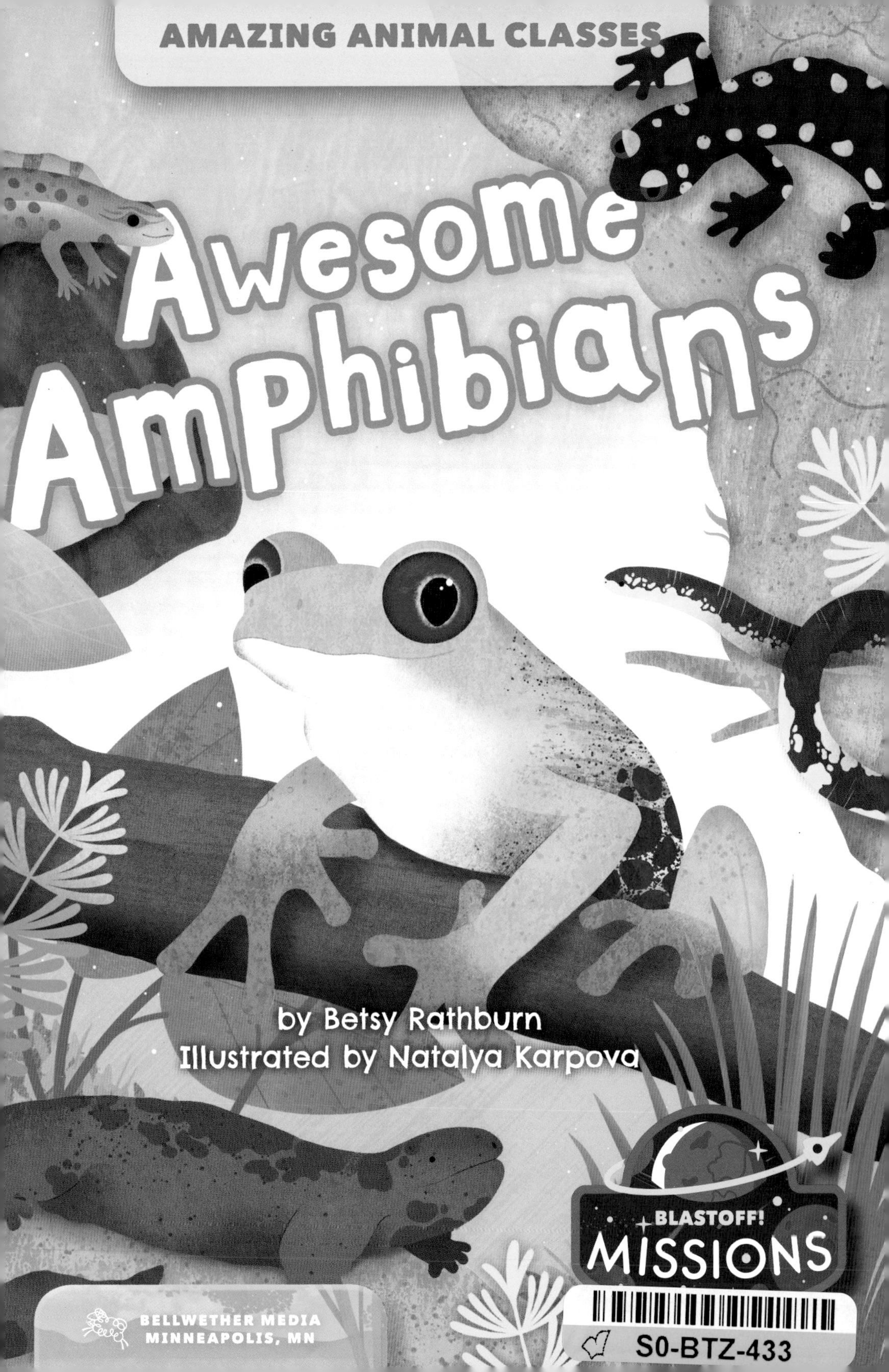

Awesome Amphibians

by Betsy Rathburn
Illustrated by Natalya Karpova

BLASTOFF! MISSIONS

BELLWETHER MEDIA
MINNEAPOLIS, MN

Blastoff! Missions takes you on a learning adventure! Colorful illustrations and exciting narratives highlight cool facts about our world and beyond. Read the mission goals and follow the narrative to gain knowledge, build reading skills, and have fun!

Traditional Nonfiction

BLASTOFF! Beginners

BLASTOFF! DISCOVERY

BLASTOFF! MISSIONS

Narrative Nonfiction

Blastoff! Universe

MISSION GOALS

- FIND YOUR SIGHT WORDS IN THE BOOK.
- IDENTIFY THE TRAITS OF AMPHIBIANS.
- THINK OF QUESTIONS TO ASK WHILE YOU READ.

This edition first published in 2023 by Bellwether Media, Inc.

Library of Congress Cataloging-in-Publication Data

Names: Rathburn, Betsy, author.
Title: Awesome amphibians / by Betsy Rathburn.
Description: Minneapolis, MN : Bellwether Media, Inc., 2023. | Series: Blastoff! Missions. Amazing animal classes | Includes bibliographical references and index. | Audience: Ages 5-8 | Audience: Grades 2-3 | Summary: "Vibrant illustrations accompany information about amphibians. The narrative nonfiction text is intended for students in kindergarten through third grade"-- Provided by publisher.
Identifiers: LCCN 2022020220 (print) | LCCN 2022020221 (ebook) | ISBN 9781644876473 (library binding) | ISBN 9781648348310 (paperback) | ISBN 9781648346934 (ebook)
Subjects: LCSH: Amphibians--Juvenile literature.
Classification: LCC QL644.2 .R378 2023 (print) | LCC QL644.2 (ebook) | DDC 597.8--dc23/eng/20220525
LC record available at https://lccn.loc.gov/2022020220
LC ebook record available at https://lccn.loc.gov/2022020221

Editor: Christina Leaf Designer: Andrea Schneider

Printed in the United States of America, North Mankato, MN.

Table of Contents

Amphibian Adventure

Amphibians are unusual animals! These **cold-blooded** creatures have backbones. They start life in water. As they grow up, most grow **limbs** to walk on land!

Let's go on an amphibian adventure!

Slimy Swimmers

Here we are near a European river. There is a smooth newt coming out of the water. It has slimy skin!

That yellow-bellied toad looks bumpy. But it is slimy, too!
yellow-bellied toad
JIMMY SAYS
Amphibians can breathe through their skin when wet. Many make a slimy coating called mucus. It keeps their skin wet!

Forest Friends

Let's explore this North American forest. See the spotted salamander on that rock? It keeps cool in the shade.

Look there! An American bullfrog is hopping near the river.

Welcome to Central America! This red-eyed tree frog climbs through the **rain forest**. Its special toes cling to leaves!

It has special eyes, too. The red color scares away **predators** such as toucans!

red-eyed
tree frog

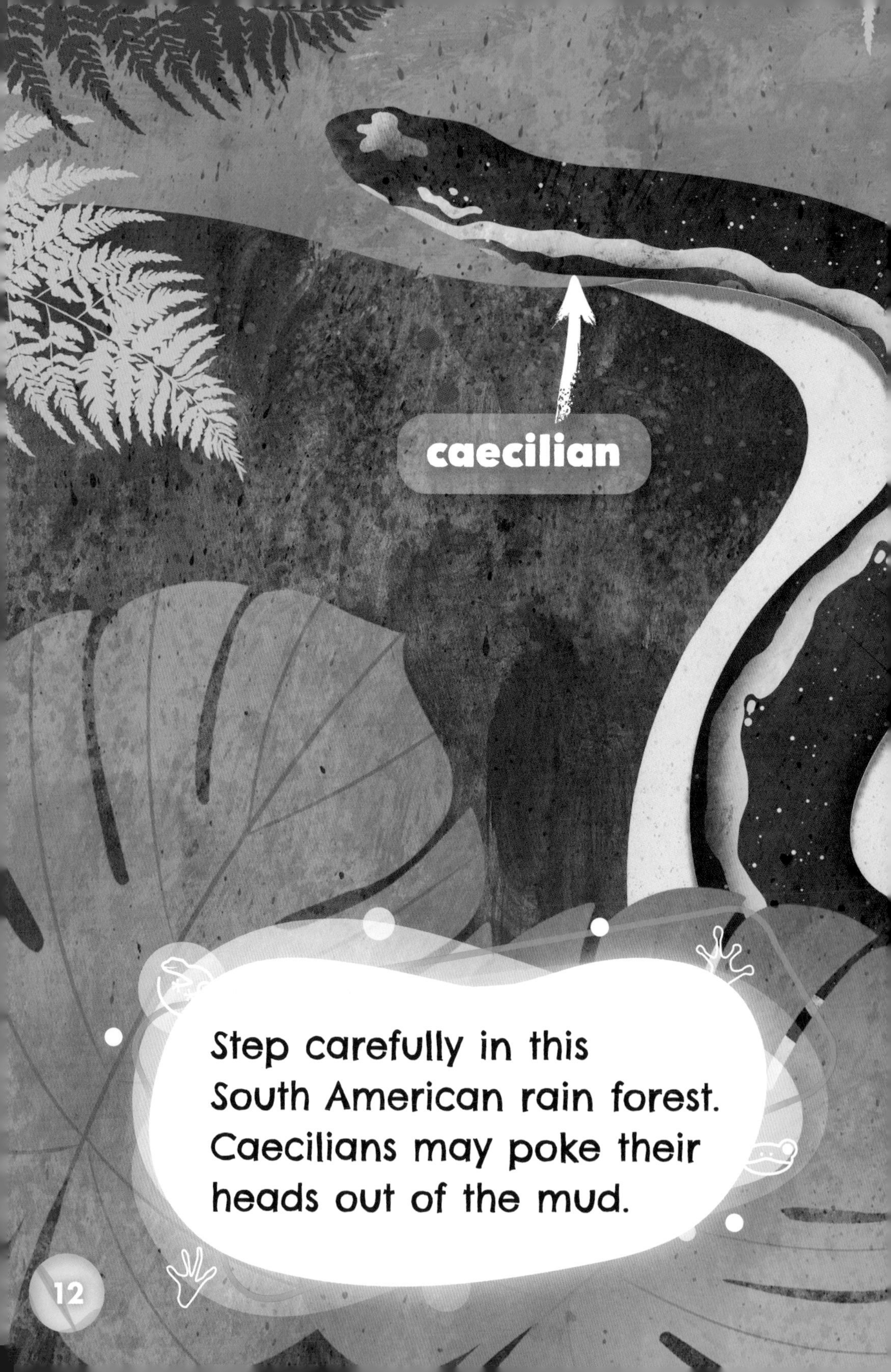

Step carefully in this South American rain forest. Caecilians may poke their heads out of the mud.

Most of these legless animals live underground. They **burrow** for worms to eat.

Those northern corroboree frogs stand out in Australia's mountain forests. But do not get too close. Their stripes mean stay away.

Predators avoid these frogs. They make their own **poison**!

northern corroboree frog

World's Largest

Here in Asia, we are looking for the world's biggest amphibian! That Chinese giant salamander hides underwater.

Chinese giant salamander

Asian toads

It waits for food to come near.
Those Asian toads narrowly escaped!

Now let's hop over to West Africa. Here, a huge goliath frog leaps after food.
goliath frog

Like most amphibians, it lays eggs in water. But unlike others, it builds its own ponds for eggs!
JIMMY SAYS
Goliath frogs are the world's largest frogs. They can weigh over 7 pounds (3 kilograms)!

We saw many amphibians on our trip. They crawled on forest floors and swam in lakes and rivers.

Be on the lookout for more of these awesome creatures!

Amphibian Facts

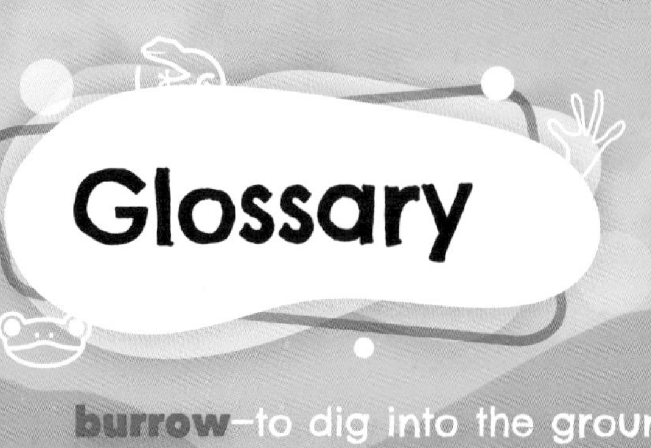

Glossary

burrow–to dig into the ground

cold-blooded–having a body temperature that changes with the outside temperature

limbs–arms and legs

poison–a substance that can cause sickness or death when touched or eaten

predators–animals that hunt other animals for food

rain forest–a thick, green forest that receives a lot of rain

To Learn More

AT THE LIBRARY

Hansen, Grace. *Goliath Frogs.* Minneapolis, Minn.: Abdo Kids, 2019.

Jaycox, Jaclyn. *Unusual Life Cycles of Amphibians.* North Mankato, Minn.: Capstone, 2021.

Kenney, Karen Latchana. *Rain Forests.* Minneapolis, Minn.: Bellwether Media, 2022.

ON THE WEB

FACTSURFER

Factsurfer.com gives you a safe, fun way to find more information.

1. Go to www.factsurfer.com.
2. Enter "awesome amphibians" into the search box and click 🔍.
3. Select your book cover to see a list of related content.

BEYOND THE MISSION

> CAN YOU THINK OF ANOTHER AMPHIBIAN THAT WAS NOT IN THE BOOK?

> DRAW A PICTURE OF YOUR FAVORITE AMPHIBIAN FROM THE BOOK.

> DO YOU THINK IT WOULD BE EASY TO HAVE AN AMPHIBIAN FOR A PET? WHY OR WHY NOT?

Index